T0224590

essentials

Springer essentials

Springer essentials provide up-to-date knowledge in a concentrated form. They aim to deliver the essence of what counts as "state-of-the-art" in the current academic discussion or in practice. With their quick, uncomplicated and comprehensible information, *essentials* provide:

- an introduction to a current issue within your field of expertis
- an introduction to a new topic of interest
- an insight, in order to be able to join in the discussion on a particular topic

Available in electronic and printed format, the books present expert knowledge from Springer specialist authors in a compact form. They are particularly suitable for use as eBooks on tablet PCs, eBook readers and smartphones. *Springer essentials* form modules of knowledge from the areas economics, social sciences and humanities, technology and natural sciences, as well as from medicine, psychology and health professions, written by renowned Springer-authors across many disciplines.

More information about this subseries at https://www.springer.com/series/16761

Petra Schling

The Sense of Taste

Of genes, molecules and the
fascinating biology of one of the
most fundamental senses

Petra Schling
Biochemie-Zentrum
Universität Heidelberg
Heidelberg, Germany

ISSN 2197-6708   ISSN 2197-6716 (electronic)
essentials
ISSN 2731-3107   ISSN 2731-3115 (electronic)
Springer essentials
ISBN 978-3-658-32232-8   ISBN 978-3-658-32233-5 (eBook)
https://doi.org/10.1007/978-3-658-32233-5

Responsible Editor: Sarah Koch
This Springer imprint is published by the registered company Springer Fachmedien Wiesbaden GmbH part of Springer Nature.
The registered company address is: Abraham-Lincoln-Str. 46, 65189 Wiesbaden, Germany

What You Can Find In This *essential*

- The biological definition of taste and an overview of the cells and molecules involved
- Molecules, receptors, and signaling pathways to the individual flavors and trigeminal stimuli
- The importance of taste for life and survival
- Deceptions of taste from nature and the food laboratory
- A look at the latest findings on the importance of taste receptors away from the mouth

Preface

This *essential* is based on lectures for students of biology and medicine and at children's universities, which I have been allowed to give for about 10 years. Of course, this text is much more detailed than a 2 h lecture could ever be. Please do not be put off by the many abbreviations. You only need to know them if you are more intensively interested in one of the molecules. The information presented here is not from my own research but has all been compiled from publications by others. You can find them in the bibliography. My thanks go to the authors of these articles, without them this *essential* would never have been possible. However, the current scientific understanding of how things are can only ever be a snapshot. Scientific "facts" are constantly being reviewed by critical people and often refuted by them. What today is regarded as certain knowledge may be outdated tomorrow. This *essential* is intended to give you an overview of the current state of research into taste. However, its statements are not set in stone and certainly carry my very personal subjective note. I would, therefore, be pleased to receive any comments you may have on the text. Only through your critical view can a further edition come still closer to the truth.

The topic of taste is of course closely linked to nutrition. And when it comes to the topic of nutrition, there are often fierce discussions about good and evil, right and wrong, far removed from scientific findings. In this context, food is often an important means of forming identity and distinguishing oneself from others (Klotter 2016). In this *essential,* however, taste and thus also nutrition should be considered from the biological necessity to distinguish essential food components from toxins. Chemistry is not a bad thing and "organic" is not a good thing. Whether a molecule was produced in a chemical laboratory or by a living being, whether it was purified (crystal sugar) or not ("natural" sugar in fruit juice), changes nothing about the molecule and its effect. The only criterion for "bio" is that the molecule has been synthesized by a living being. However, nature has "invented" the strongest poisons of all, and plants and animals do not want to be eaten by us. "Chemistry" in food has made it safer (DKFZ 2016) and is causing a continous fall in new cases of stomach cancer and mortality rates (Robert Koch Institute 2017). Genetic engineering and other genetic modifications have turned toxic plants into edible food. There is no doubt that the food industry has not only produced good things. For example, the greening of canned vegetables was banned as early as 1887 because of the known toxicity of the copper sulfate used. In 1928 it was permitted again so that the domestic vegetables did not look so colorless next to the competition from abroad (Goldstein 1954). And the poisonous copper sulfate is still one of the most widely used pesticides in organic farming today, although more specific pesticides that are gentle on beneficial organisms are available from chemical laboratories (Kaufmann 2016).

In this sense: Let your nose—no, your taste buds!—show you around and enjoy it!

Petra Schling

Contents

Taste From a Biological Perspective

1

Summary

The German word "Geschmack," like the English word "taste," describes in everyday language various contents, from "fashion taste" to "tasteless choice of words" to the "taste of chocolate." Here, however, it should not be about the description of social norms, but about a biological sense. So the third example hits the nail on the head, doesn't it?

1.1 Differentiation of Taste From Other Sensory Impressions

The "taste of chocolate" also contains at least three, sometimes four different sensory impressions, which only complement each other in the brain when they occur and are perceived together to form the unique chocolate taste (see Fig. 1.1a):

- The texture conveyed by tactile sensors in combination with, for example, sucking movements;
- The smell of the molecules released into the gaseous phase by the heat and by sucking or chewing, and which reach the nose mainly via the pharynx ("back of the throat");
- The taste in the genuine sense, which is recognized by the taste buds and the taste cells they contain; and
- Pain stimuli to which free nerve endings in the oral cavity react and which are quite consciously integrated into many a chocolate.

© Springer Fachmedien Wiesbaden GmbH, part of Springer Nature 2021   1
P. Schling, *The Sense of Taste,* essentials,
https://doi.org/10.1007/978-3-658-32233-5_1

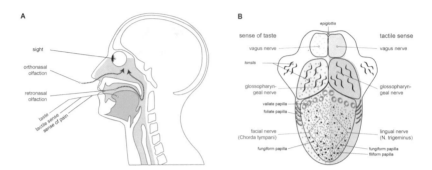

Fig. 1.1 Anatomical position of the senses in the head. **a** Overview. **b** Tongue. (The different tastes can be recognized throughout the tongue, from the vallate and foliate papillae to the tip of the tongue. However, the sensitivity to touch decreases from the front tip of the tongue to the back). *Source* Own representation

So what we usually refer to as taste is what the English word "flavor" means: A combination of smell, taste, tactile sensation, and sometimes pain. However, these four sensory impressions influence appetite and eating behavior in different ways (Boesvels and de Graaf 2017). It is, therefore, worth trying to separate these senses. If we borrow words from Latin and Greek, our perceptions of food can be distinguished semantically exactly in gustatory (tasting in the true sense), olfactory (smelling), epicritical (sense of touch for contact and exploration), and protopathic (temperature and pain).

More difficult is an experimental differentiation in real life. The influence of the smell can be eliminated quite easily by putting on a nose clip. It is much more difficult to trick the influence of touch: In this case, comparative dishes must be prepared in such a way that they do not differ in texture. And if pain stimuli are not to influence the experiment, the corresponding nerve (trigeminal parts of the lingual nerve) would have to be inhibited with local anesthetics. This can hardly be implemented experimentally since the trigeminal nerve fibers in the tongue's area also attach to parts of the facial nerve, which transmit the actual taste information of the front two-thirds of the tongue (see Fig. 1.1b). If local anesthetics inhibit the pain stimulus, it also restricts the sense of taste. Another variant is, therefore, the desensitization of pain fibers by chemical overstimulation with capsaicin (see also further down in Sect. 2.4). Capsaicin only numbs the nerves that have the corresponding heat receptors and leaves the taste buds unaffected.

The experiments conducted so far seem to allow the following conclusion to be drawn: If the test persons are asked whether they like the taste of food; then

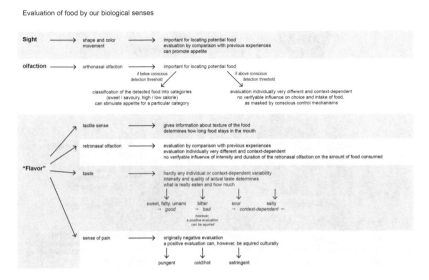

Fig. 1.2 Influence of different sensory modalities on our search for food (appetite), food selection, and food intake. *Source* Own presentation

appearance, smell, texture, and one or the other well-placed pain stimulus are decisive. If, however, it is measured in the experiment how much of the food is actually eaten, then only the gustatory perception, that is, the taste in the narrow sense is decisive (see Fig. 1.2).

1.2   Papillae and Buds: A Closer Look at the Tongue

Taste is created when chemical stimuli, that is certain molecules, bind to the receptors on specialized sensory cells in the tongue. These taste cells are located in so-called taste buds (see Fig. 1.3b), in which 50–100 such cells are arranged like the segments of an orange (Chaudhari and Roper 2010). On the tongue, the taste buds are located in the papillae (leaf, wall- and fungal papillae) (see Fig. 1.1b and 1.3a). However, in other parts of the mouth, taste buds can also be found freely in the mucous membrane. Each taste bud has a similar repertoire of different gustatory cells so that the taste sensations at the different parts of the tongue do not differ significantly. The tip and edge of the tongue are more sensitive to all tastes because of the papillae that are more abundant there.

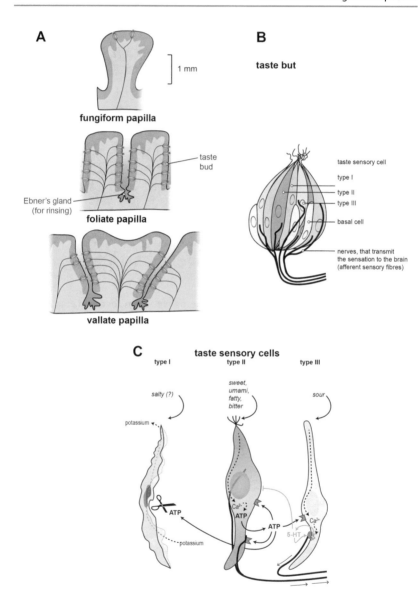

Fig. 1.3 **a** Taste papillae of the tongue. **b** Taste bud. **c** Different types of taste cells. *Source* Own representation

The elongated cells of the taste buds have direct contact with the saliva and thus with the outside world. They have to resist fluctuations in temperature, pH, and salinity and sometimes have contact with toxic substances. Their lifetime is, therefore, limited and they are continuously regenerated from the basal cells. In addition to the basal cells, three different types of gustatory cells are distinguished, which were called (not very imaginative) types I, II, and III (see Fig. 1.3c).

Type I cells are the most common cells in a taste bud and seem to be mainly busy cleaning up. They remove all the substances from the intercellular spaces that accumulate during active taste perception: Especially potassium ions and neurotransmitters. However, it is also possible that it is precisely these type I cells that are responsible for the salt taste.

Type II cells are classical sensory cells that express receptors for one of the following tastes: Sweet, umami, fatty, or bitter. The receptors are located on cell protuberances at their uppermost tip, which protrudes from the taste bud like a shock of hair. In other organs, these cells are, therefore, sometimes called "tuft cells." If they are stimulated with the corresponding taste molecules, the concentration of calcium ions (Ca^{2+}) increases in their interior and they release the messenger substance adenosine triphosphate (ATP) through pores in the membrane. This ATP not only acts back on the sensory cells itself but also activates the associated nerves and neighboring sensory cells and is finally degraded by the type I cells.

Type III cells resemble real nerve cells, in that they release vesicles with the messenger substances serotonin (5-HT) and noradrenalin after stimulation and a corresponding increase in calcium concentration. Serotonin acts directly at the synapse on the associated nerve fiber and can also reach neighboring type II sensory cells and inhibit their activity. Type III cells are responsible for the recognition and transmission of acid and also the tickling sensation of effervescence. At the same time, however, type III cells also react to ATP from type II cells, that is, they become co-excited by sweet, umami, fatty, and bitter.

Although individual taste cells are originally specific for only one flavor, the tight packaging within the taste bud and the manifold communication between the different cell types already complicates things at the level of the tongue. How these complex patterns of signals are encoded in the nerves and decoded in the brain is still completely unexplained.

Flavors

2

Summary

According to our definition from the first chapter, all that is a taste is what we can perceive without the involvement of nose or eyes and without the sense of touch or pain. Ideally, a cell in a taste bud should be responsible for expressing specific proteins as receptors and, after stimulation, trigger a signaling cascade that can be measured down to the specific gustatory nerves.

Some tastes are almost certain here: Sweet, umami, fatty, bitter, and acidic. Nobody would want to deny that "salty" is a real taste, but so far, neither the cell type has been reliably identified nor the signal chain understood. Water is also not yet fully understood, but probably also a candidate for a real taste. Pungent and astringent are not part of the taste, but of the perception of pain and temperature.

2.1 Our Favorites: Sweet, Umami, and Fatty—from a Cat Snitching Cake, Hummingbirds, and Greek Yogurt

Since food in our evolutionary past was usually scarce and only obtainable at considerable risk, we humans and many other animals are specialized in not spurning high-quality food when it is available to us. And what is high quality? Well, from an evolutionary point of view mainly energy dense food with the three macronutrients: Carbohydrates, fats, and proteins.

Carbohydrates are made up of building blocks of sugar and after vigorous chewing and salivation in the mouth, the sweet-tasting sugar molecules are released. It is similar with fats and proteins. We cannot taste the large molecule, but we can taste the fragments, which are released in small amounts already in the mouth in saliva by the appropriate digestive enzymes. All three tastes and also bitter (see

© Springer Fachmedien Wiesbaden GmbH, part of Springer Nature 2021
P. Schling, *The Sense of Taste,* essentials,
https://doi.org/10.1007/978-3-658-32233-5_2

below at Sect. 2.2) are registered by type II taste cells, each of which expresses the appropriate receptor for one of the molecule classes (see Fig. 2.1).

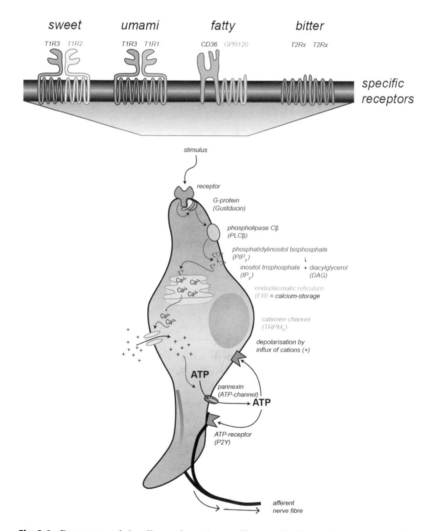

Fig. 2.1 Receptors and signaling pathway in type II taste cells. *Source* Own representation

As soon as the "right" molecule docks to the respective receptor, the latter activates the internally attached protein gustducin, which dissociates from the receptor and diffuses to another membrane protein, phospholipase C (PLC), which is activated by it. The PLC is a protein that accelerates the cleavage of the membrane lipid phosphatidylinositol-4,5-bisphosphate (PIP2) into two fragments, one of which, the inositol trisphosphate (IP3), is water-soluble and can be distributed in the cell interior. Here it meets a channel in an inner cell organelle, the so-called endoplasmic reticulum (ER). The ER is the most important intracellular calcium store of a cell. IP3 opens this channel and calcium ions flow out into the cell interior. These calcium ions in turn open a cation channel in the plasma membrane, the transient receptor potential channel member 5 (TRPM5), which now allows cations, especially sodium ions, to enter the cell from outside. The influx of positive charges electrically excites the cell and opens ATP channels that allow ATP to flow from the cell interior to the outside where it activates the associated afferent nerves, among other things. Like a series of dominos, one messenger nudges the next—with the difference that amplification factors are integrated into the signaling chain: For example, one receptor activates several gustducins, a PLC cleaves many PIP2s that open several calcium channels and release many calcium ions, etc. A few taste molecules are, therefore, sufficient to trigger an avalanche in a taste cell and to excite it.

The receptors for the taste molecules are each composed of two individual protein chains. The sweet and umami receptors even share one of these chains: Chain No. 3 of the "taste receptor type 1" (T1R3). Thus, if an animal loses the ability to produce this protein chain, the animal cannot taste sweet or umami with its usual distinctness. If one of the other type 1 chains is missing, only one taste, either sweet or umami, will be missing. The individual protein chain of a type 1 taste receptor winds its way seven times through the plasma membrane and then forms a ball and slot on the outside (i.e., where the saliva is). This spherical part looks like Pac-Man and is called Venus flytrap domain. When a taste molecule binds to this domain, it snaps shut, and the movement is transmitted through the membrane to the inside of the cell.

Sweet
Carbohydrates are produced by plants from sunlight, carbon dioxide (CO_2), and water (H_2O) and are stored for dark times (at night) and the offspring. Carbohydrates are, therefore, THE energy stores par excellence in our world and are used not only by plants but also by those who eat plants. Sugar is essential for the supply of our red blood cells and our brain, and all other organs like to use sugar as an energy source. No wonder that carbohydrates = sweets is one of our favorite flavors. Plants

"know" this and, therefore, reward us for spreading their seeds, for example, from apples or peaches, with sugar in the pulp. The seeds themselves (apple or peach seeds) we "should not" eat, of course—as they do not taste sweet, but are spoiled with bitter substances by the plant (see below at Sect. 2.2).

Cats

like to nibble—our cat sometimes steals a piece of cake from our plate. More surprising was the discovery that all cats, from tigers to domestic cats, can no longer express a functioning protein chain T1R2. All of them have no receptor for sweetness and cannot taste the amount of sugar or carbohydrates in a meal. The researchers suspect that this is a classic case of "use it or loose it." Cats have been feeding solely on prey for many millions of years, and as pure carnivores it was irrelevant whether the mouse or gazelle had eaten a little grass before. So when the mistake in the genetic material happened, it was no disadvantage. Cats, however, naturally have a very sensitive taste for umami and fat—and in the cake, besides flour and sugar, there is enough butter and eggs. This is what makes our cat snitch the cake.

Umami

Animal protein is an important and biologically high-quality source of protein for us and other carnivores and omnivores. Proteins are partially broken down in our mouth into their building blocks, the amino acids, and the most common of these, glutamate, we humans can taste. Many carnivorous animals can probably taste a larger repertoire of amino acids than we can. In order to be able to differentiate between animal and vegetable protein in terms of taste, a special feature has been developed in the umami taste, which exploits a clear difference between animal and vegetable food: Animal cells are on average much smaller. However, since each cell contains about the same amount of genetic material (Deoxyribonucleic acid = DNA), animal food has more DNA per cell than plant food. DNA is also a large molecule consisting of individual building blocks, the nucleotides. So when we eat cells, the digestive enzymes in saliva also release nucleotides—much more so in animal food than in plant food. Both amino acids, especially glutamate, and nucleotides can activate the umami receptor. However, if both occur simultaneously, the effect is synergistic rather than additive. Together they are more than the sum of both individual stimuli. Strangely enough, glutamate is often referred to as a "flavor enhancer," although it is actually the nucleotides that act as flavor enhancers. In any case, one of them enhances the umami taste of the other.

The Giant Panda

belongs to the group of carnivores, but changed its diet to purely vegetable about 4 million years ago. Today, it only eats bamboo. About 4 million years ago, its T1R1 gene, the crucial part of the umami receptor, mutated in such a way that it no longer functions. However, other herbivores, such as cows and horses, still have a functional umami receptor. What these two need their umami receptor for, or whether it is pure coincidence that the gene has not yet mutated, cannot yet be answered.

But without the classic umami receptor from T1R1 and T1R3, we can still taste Glutamate a little bit. This is done via a truncated version of the neuronal glutamate receptor (mGluR) with a shortened Venus flytrap domain. In its intact form, the neuronal receptor would be far too sensitive to the huge amounts of glutamate we ingest with food. The truncated version is, therefore, tuned down to the sensitivity required in the mouth. However, an amplification of the glutamate taste by nucleotides is not possible with the mGluR.

Hummingbirds

feed only on sweet nectar—and show what detours evolution sometimes has to take: Like cats, birds have all lost the gene for T1R2, the classic sweet receptor. Since the birds' closest relatives, the crocodiles, still have the T1R2 gene, the gene loss probably occurred in the common ancestor of all birds during the time of the dinosaurs. But then why do hummingbirds like the sweet nectar of a flower? This question could be clarified by comparing them with closely related birds (e.g., the common swift). Swifts feed exclusively on insects and cannot taste "sweet." Hummingbirds, in comparison with the swift relatives, have their umami receptor dimer of T1R1 and T1R3 elaborately converted (Baldwin et al. 2014). The hummingbird receptor now recognizes sugar instead of amino acids. For this purpose, the receptor had to be modified at 19 sites. Each individual exchange happened through a random mutation in the corresponding gene and had to prove positive for many generations of hummingbirds. I wonder how many birds hatched along this evolutionary path whose taste receptor could neither taste umami nor sweet?
Hummingbirds may be the best researched, but they are not the only ones. Nectar birds as well can taste sugar. They occupy a similar ecological niche, but they belong to the sparrow family and are, therefore, hardly related to the hummingbirds. It will be exciting to find out which way the ancestors of the nectar birds found to be able to taste sweet again.

Fats

Fat contains the most calories per gram of food. In times when food is scarce, a fatty meal can mean survival. And so we know from our own experience: Fatty foods make you happy. Think of chocolate, chips, pizza or even cake. The fat we like here are the so-called triacylglycerides, that is, three fatty acids combined with one molecule of glycerine. According to everything we know so far about the perception of "fats," we cannot taste triacylglycerides. However, they influence the texture of the food and give a positive, creamy mouthfeel via our sense of touch. The taste is triggered by free fatty acids which are released in the mouth by an enzyme in saliva, the lingual lipase (see Fig. 2.2).

Fat content in dairy products

For a long time, **"fat"** was considered to be a mouthfeel—not a taste. And even if the consistency—perceived through our sense of touch—is important, a cream yogurt with 10% fat content simply tastes better to us than a low-fat dairy product, no matter how creamy it may appear. In a test for the gourmet magazine "Falstaff," the sensory experts and cheese makers came to the conclusion: "The 'fat-is-good theory' has been confirmed in practice, because the best three products have the highest fat content." (Starz 2017). Real Greek yogurt also has 10% fat, as it is virtually yogurt concentrate. In the USA, on the other hand, "Greek yogurt" does not have much to do with Greek

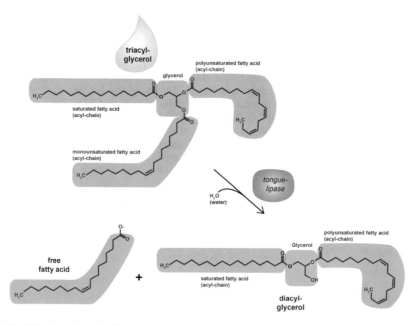

Fig. 2.2 Reaction of the lingual lipase: Release of individual fatty acids during eating. *Source* Own representation

yogurt and is rather low fat (0– max. 5% fat). Nevertheless, it has conquered over 30% of the market there (Chandan 2013). Is it possible that in the US market, unlike in Europe, the focus when buying yogurt is not on taste but on slimness?

Two proteins are found on the corresponding type II gustatory cells, which are potential molecular receptors for the free fatty acids: Only GPR 120 (G-protein-coupled receptor) can activate the classical signaling pathway via gustducin (see Fig. 2.1). However, CD 36 is at least as important: Mice without CD 36 cannot taste fatty acids and in humans, the individual differences in fat perception are due to differences in CD 36. So far, the exact function of CD 36 can only be speculated about. CD 36 binds to fatty acids much more strongly than GPR 120. It is possible that it collects and concentrates fatty acids in the surroundings of GPR 120, which would not actually become active with the normal concentrations in the mouth. Or they may function as a couple: CD 36 binds the fatty acid, triggers GPR 120 and this then signals to the cell's interior.

We therefore find the signals of the taste sensory cells type II with fat receptor to be delicious if the fat in the food is mainly present as triacylglycerides and thus additionally gives us a pleasant mouthfeel via its texture. Free fatty acids alone, for example, dripped on the tongue as a test solution, on the other hand, are perceived as unpleasant and associated with rancid, spoiled food. In this scenario, we can probably smell the fatty acids and the sense of taste is not matched by the pleasant sense of touch. The classic soap (curd or soft soap) consists of free fatty acids and salt. And who likes the taste of soap in the mouth? Ingested as triacylglycerides and released by the lingual lipase during chewing, fatty acids taste delicious. If they are already present in food, they repel us (Reed and Xia 2015).

Sharks

love it fatty—they are usually very intelligent and have highly developed senses. What they do not know, they curiously explore (with due caution). Since they have no hands, they "grasp" it with their mouth. This behavior is best studied in the great white shark (Martin 2010). In the living shark the teeth are flexibly suspended and highly, sensitively, innervated. By means of this sense of touch, a great white shark can learn precisely how an object feels through a test bite—for example, whether there is a thick layer of fat under the skin. Unfortunately, the sense of taste in sharks has hardly been investigated to date, but between the rows of their teeth there are particularly many taste buds that help to analyze the test object. They are already fully developed in embryos, long before the teeth. It is, therefore, likely that sharks also taste and learn in the womb or in the egg what their mother prefers to eat. From observations of the great white shark, a preference for "fatty" and "umami" can be inferred, in that order. Especially the great white shark has to take care that his food contains a lot of fat, because he is warm-blooded and has a very slow digestion. In order to maintain its body temperature in cold water, it is not sufficient to fill its stomach with lean meat. But sharks are flexible and, of course, do a benefit/risk assessment. If the potential food is harmless, like a dead sheep carcass thrown overboard from a ship, then the shark can accept a little less fat. However, if the food beats or bites around during the test bite and then does not even taste sufficiently fatty, the shark prefers to leave it at the one test bite and swim away (Smallwood 2016).

Whales and dolphins

only have a few taste buds when they are young. As young animals, they can distinguish fish taste from water, as adults not anymore. However, it is unclear how they do this, since the genes for their sweet, umami, and bitter receptor proteins are so severely mutated that they are useless. Perhaps they can taste salty (Bouchard et al. 2017). Probably they do not need a good sense of taste because they swallow their prey whole, anyway. But how especially the adult animals know that they swallow an edible fish and not a poisonous one remains their secret.

2.2   A Well-meant Warning: Bitter—Eat and Be Eaten

Bitter = poison. The bitter taste warns us about poisonous and dangerous things in our food and we do well to respond to this warning accordingly. In 2015 there was even a death in Germany because a man ate zucchini even though it tasted extremely bitter to him. Such a tragic outcome is very rare, however, as most animals (including humans) usually spit out bitter foods immediately. It is, therefore, important for survival that we can taste the many different poisons that occur in nature. Therefore, we humans have at least 25 different bitter receptors. They belong to the "taste" receptors, but do not have a Venus flytrap domain like the sweet and umami receptors, and are therefore assigned to group 2 (abbreviation: T2R, see Fig. 2.1). Each T2R can bind to several bitter substances (toxins) and each type 2 taste cell, which is responsible for the bitter taste, carries several different T2R on its surface. So we cannot distinguish different toxins, but we can trust that one of the many receptors on one of the many bitter taste cells will already give an alarm. The signaling cascade that is triggered is identical to that of the other type 2 flavors (sweet, umami, fatty, see Sect. 2.1). With so many different bitter substances and at least these 25 different receptors, it will take some time before the researchers have found all natural ligands, that is, toxins, for each receptor. T2R number 16 and number 38 are the best studied.

T2R16

binds to β glucopyranosides, among which are the particularly toxic cyanogenic glycosides such as amygdalin from bitter almonds and apricot kernels (see Fig. 2.3). There are two functional genetic variants of T2R16: An ancient one with lysine as the amino acid at position 172 (K172), which binds the glycosides only weakly and is called the "non-taster" variant, and a newer one, which probably originated in East Africa about a million years ago and carries asparagine as the amino acid at this position (N172). N172 is the "taster" variant of T2R16. Since β glucopyranosides are very widespread (over 2,500 plants and insects synthesize them as protection against predators) and many of them are very toxic, it is not surprising that people have had advantages with N172. Today, this variant of the gene dominates with over 98% of the world population. Only in Africa, especially West Africa, more than 10% of people still live with the K172 variant, which can hardly taste β glucopyranosides.

T2R38

binds the two synthetic bitter substances phenylthiocarbamide (PTC) and 6-n-propylthiouracil (PROP) (see Fig. 2.3), but probably arose to detect natural glucosinolates, for example, from cabbage and broccoli. These secondary plant compounds give representatives of the crucifer family such as radish, horseradish, mustard, cress, and cabbage their somewhat pungent and bitter taste. These plants

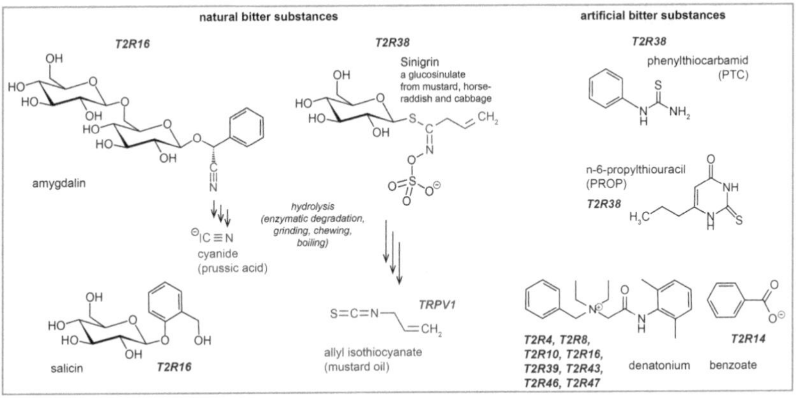

Fig. 2.3 Bitter substances and bitter receptors that they activate and toxins they release (TRPV1 = transient receptor potential channel/vanilloid receptor type 1 = "pungent receptor"). *Source* Own representation

form glucosinolates and, in separate cell compartments, an enzyme that splits isothiocyanates (mustard oils) from the glucosinulates. The enzymatic degradation
starts when the cell is injured when the plant is nibbled by a predator. Isothiocyanates protect plants due to their pungent taste (see Sect. 2.4) and are toxic (especially
to insects). In humans they irritate the skin, respiratory tract, and gastrointestinal
tract and may damage the thyroid gland. The T2R38 also has a "non-taster" variant
(PAV for **p**roline, **a**lanine, **v**aline) and a "taster" variant (AVI for **a**lanine, **v**aline,
isoleucine). Worldwide, both variants have approximately the same frequency, resul-
ting in about 25% "super-tasters" (PAV/PAV), 25% "non-tasters" (AVI/AVI) and
50% in between (PAV/AVI). Again, the genetic variability originated in Africa about
1 million years ago and it is completely unclear why the "non-taster" variant is so
persistent. Research has gained new momentum since T2R38 is now also believed
to play an important role in the innate immune system of the lungs and intestines
(see Sects. 3.1 and 3.3).

Denatonium Benzoate
Was developed as an artificial bitter substance which, regardless of genetic variability, is perceived by every human being (and also many animals) as extremely bitter.
In humans, denatonium alone activates eight different bitter receptors and benzoate
activates at least one more. Denatonium benzoate is thus the most bitter substance
known to date and is used to denature alcohol and detergents, for example.

Eat and Be Eaten
To survive, we must eat. But since we humans cannot photosynthesize, we eat
other living things. But they do not want to be eaten by us. While animals can
usually run away or defend themselves with claws and teeth against our attacks,
many plants resort to poison. Think, for example, of the prussic acid released from
the cyanogenic glycopyranosides of almonds, apricots, and apple stones, or the
isothiocyanates from cabbage. Our vegetable foods are, therefore, often toxic by
nature and can only be eaten in very small quantities. Some of these food poisons can
be destroyed by heating or cooking, so that otherwise inedible plants can now be used
as food. One example is the poison phasin, which is found in pulses like the garden
bean, or the ricin from the seeds of the miracle tree. While even a few raw beans
can cause symptoms of poisoning, cooked bean soup is harmless. However, most
poisons cannot be destroyed by heating, for example, solanine from the green parts
of potato and tomato plants or cucurbitacin in pumpkin plants. Some become even
more inedible by heating, such as the glucopyranosides and glucosinolates shown
in Fig. 2.3. Here, our ancestors have bred varieties over generations that are hardly
able to produce the bitter toxin due to genetic defects. These cultivars are at our

and all the other predators mercy without protection. Their cultivation is, therefore, only successful through the use of pesticides. "Old varieties" are, therefore, usually more resistant to pests, but are not necessarily suitable for consumption. When these cultivars did not yet exist, people with a weakly developed bitter taste might even have had advantages. They could use a more varied range of plant foods, even if this certainly caused symptoms of poisoning every now and then. However, when new, unknown sources of food had to be found, those who were particularly sensitive to toxic tastes were in demand.

Bitter Medicine

Drugs intervene massively in our body's own functions. This may help with an illness, but is toxic for a healthy organism. Since many drugs are structurally derived from natural toxins, they also taste bitter. A second reason for bitter pills: If the active ingredient itself does not taste bitter, a bitter substance is sometimes added to the colored pills specially so that small children cannot confuse them with colored sweets. For the confusion surrounding bitter medicine, see also Sect. 3.4.

2.3   The Side Stage: Salty, Acidic, and Water—Sparkling Water and Unripe Fruit

Salty

Our ancestors came from the sea. This is where the first unicellular organisms developed and their entire mode of functioning was based on the salts and their concentrations in seawater. For the multicellular organisms that developed from this and for us as today's land animals, it was no longer possible to change anything fundamentally. In the liquids that wash around our cells, especially the salt cations are still present in almost the same ratio as in seawater: 94 sodium ions for every 3 potassium and 2 calcium ions (Neukamm 2014). So we still carry the sea within us and that is the only way we function. But what we eat and drink no longer has the right cation composition. In drinking water there is more calcium than sodium, common salt is pure sodium chloride and orange juice contains far too much potassium compared to sodium. Our internal organs, above all the intestines and kidneys, must constantly perform at a high level in order to absorb the right ions from such a wide variety of foods and release the excess ions. But none of this helps if the supply of ions is not right in principle. Whether salty is pleasant to taste is, therefore, determined by our body's need for the right ion. And we have to be able to distinguish sodium, of which we need a lot, from the other cations (potassium, calcium, magnesium), of which we only need very small amounts. Although the taste of salt is so important and fascinating and research has been going on for over 150 years, unfortunately very little is known about the molecular conditions on the tongue. The difficulties begin with the fact that the responsible cell type has not yet been identified. All three cell types and even the basal cells come into consideration. And the proteins involved that have been identified so far do not yet yield a continuous signaling chain. Nevertheless, I would like to put together the few pieces of the puzzle here, which hopefully in the future can finally be connected to form a complete picture of the taste of salt:

The **amiloride sensitive salt taste** is physiologically seen as the positively felt salt taste of solutions with sodium ions in the concentration range of our blood (140 mM). Amiloride inhibits specifically an ion channel, the epithelial sodium channel (EnaC), and this channel could also be detected in cells that are amiloride-sensitive to sodium ions. Cells that mediate the amiloride-sensitive salty taste seem to be mainly located in the fungal papillae.

Why do tears taste salty, but saliva and sweat do not?

Our saliva, which washes around the upper ends of the taste cells all the time, is logically neutral in taste for us. We can only taste changes in the concentrations of the taste molecules. So if saliva had the same ion composition as our blood, that is, about 140mM sodium ions, the positive amiloride-sensitive salt taste between 10 and 200mM sodium would not be possible. This is where the salivary glands come into play, which first filter out the liquid of the blood including all its ions, but then transport mostly sodium and chloride ions back into the blood. In resting saliva, only 5mM sodium ions are present and any increase in this concentration can therefore trigger a taste. The lacrimal glands do not do this work and so tears taste salty. Fresh sweat, where the water has not yet evaporated, also tastes very little salty, because the sweat glands also transport back sodium and chloride. Here it is not a question of taste, but of avoiding too much salt loss through sweating.

The **amiloride-insensitive salt taste** is physiologically the negatively perceived salt taste that we experience with oversalted foods (too much sodium) and also with other cations. The responsible cells are located in taste buds in the back of the tongue, for example, in the wall papilla, and have no ENaC. There is experimental evidence that both a subgroup of bitter taste cells (type II) and additionally of acidic taste cells (type III) are involved in this component of salt taste (Lewandowski et al. 2016). While an oversalted food simultaneously triggers the amiloride-sensitive (positive) salty taste via the ENaC cells and the amiloride-insensitive (negative) taste via the bitter/acidic taste cells, other cations such as calcium and potassium only trigger the amiloride-insensitive taste and test persons report a bitter/disgusting rather than a salty taste.

Kokumi

Calcium ions are not quantitatively decisive in maintaining the salt content of extra- or intracellular fluids, but in the form of hydroxyapatite $Ca_5(PO_4)_3OH$ they are the main component of our bones and teeth. Each person thus carries about 1 kg of calcium in himself. Only 0.1% of our calcium is used for other purposes, the most important being the excitability of cells, as already explained in the signaling pathway of type II taste cells. For this purpose, the calcium concentration inside and outside the cell must be adjusted very precisely and kept constant. When there is a calcium deficiency, calcium is mobilized from the bones, which gradually become more porous. So if calcium in our food was exclusively associated with a negative taste experience, we would probably avoid food containing calcium, with corre- spondingly negative effects on bones and teeth. The solution seems to be a taste sensation called Kokumi. Presumably it is certain type II (or also III) taste sensory cells that express a calcium-sensitive receptor (CaSR). This CaSR belongs to the same family of 7 transmembrane receptors as the sweet, umami, and bitter receptors and also binds its ligands to a large Venus flytrap domain. The signal transduction also proceeds similarly via phospholipase C and outflow of calcium from the cell's internal reservoir. Although the taste substance (calcium), the receptor and the signa- ling chain have been clarified, Kokumi is not a "real" taste for a very simple reason: Kokumi substances, including calcium, are tasteless by this route. However, they enhance the positive taste of sweet and umami, and the greater the calcium defi- ciency, the stronger the taste. Kokumi is thereby a positive mouthfeel and lasting taste richness of those foods, which contain a ligand for the CaSR in addition to sweet or umami substances. This explains why calcium ions alone always taste awful, whereas calcium-rich foods like cheese or bananas do not.

Acidic, Water, and CO_2

These three apparently completely different taste qualities are all mediated by one and the same taste cell and are based on the sensitivity of these cells to changes in the concentration ratio of the two "water ions" H_3O^+ and OH^-. More H_3O^+ than OH^- means acidic, more OH^- than H_3O^+ basic. The pH value is a measure of the concentration of H_3O^+ ions and although it is kept extremely constant within our body, it may fluctuate more strongly on our surfaces (mouth, stomach, intestine, bladder, skin). In saliva, the salivary glands set a slightly alkaline pH value, especially when we eat, because otherwise our teeth will dissolve due to the acids from the food (caries). The pH value of saliva is mainly maintained by the anion hydrogen carbonate (HCO_3^-), which decomposes with acids to carbon dioxide (CO_2) and water, neutralizing the acid:

$HCO_3^- + H_3O^+ \rightarrow 2\,H_2O + CO_2$

However, the reaction is freely reversible, so that more water or CO_2 dissolved in it also causes the back reaction and causes the pH value to become acidic:

$2\,H_2O + CO_2 \rightarrow HCO_3^- + H_3O^+$

However, without an enzyme, the forward and backward reaction takes place very slowly. This is why sparkling water is not actually acidic, but only contains a lot of physically dissolved CO_2. The acidic taste cells, however, carry the enzyme carbonic anhydrase on their surface, which accelerates the back and forth reaction extremely. In the vicinity of an acid taste cell, any change in the concentration of one of the molecules leads to a change in pH. The sensory cells, therefore, only have to react to pH fluctuations in order to be able to taste all three molecules.

A large surplus of H_3O^+ is perceived as negative and contributes to the rejection of unripe or spoiled food. However, the **acidic taste,** which is mediated by taste cells in the taste buds of the tongue, does not seem to be responsible for this. Acid can also taste good to us through the taste sensory cells: Together with a lot of sugar, which indicates that the fruit is ripe, as an indication of water when we are thirsty, or just tickling locally, as with fizzy drinks. The unpleasant acidic "taste," as well as pungent and astringent (see Sect. 2.4), is transmitted by pain sensors in the mouth (Zocchi et al. 2017).

Even though the actual receptor protein for the acidic taste has still not been identified, the cell type has at least been discovered. It is a subgroup of type III taste cells in the taste buds that express a specific ion channel (PKD2L1). However, this channel itself is not important for the acidic taste, but serves as a good marker for finding the acidic taste cells among the many different type III cells in a taste bud. Investigations on these cells suggest that acid taste cells are only activated by protons (H_3O^+) inside the cell. Protons can themselves flow through a channel in the cell membrane, or they can piggyback on weak, organic acids such as citric acid or acetic acid across the cell membrane. Because the second process is more efficient, organic acids taste much more acidic than strong hydrochloric or phosphoric acid at the same pH value.

It is enormously important that we can also taste **water.** The amount of water in our body ensures that we have sufficient blood pressure and also the right concentration of salt in our body fluids. In mammals like us humans, there are specific nerves in the brain that measure the internal water content of the body and regulate the sensation of thirst accordingly. Not every liquid that looks like water, has the same viscosity and smells like nothing, is really water. The distinction between, for example, thin-bodied silicone oils and real water can only be made on the basis of taste. Mice that genetically lack the acid taste cells also drink silicone oil (Zocchi et al. 2017). The dilution of saliva with water, on the other hand, leads to an increased local formation of H_3O^+ with the help of carboanhydrase and thus to an activation of the acid cells. And when these are activated, positive feedback with the brain leads to increased drinking behavior until the sensors in the brain report that the body is once again sufficiently supplied with water.

The **taste of CO_2** in sparkling water, lemonades, and sparkling wine is based, like the taste of water, on the acidic taste cells and the carboanhydrase attached to them. As a result, these drinks always taste slightly acidic. The fact that the feeling of CO_2 is also tickling and pungent is, as with classic acids, again due to the pain sensors.

2.4   The Trigeminal Stimuli: Pungent, Sharp, and Astringent—About Chilies and Wine

Pain serves to warn us of harmful influences. These are objectively registered by free nerve endings in the tissue, so-called nociceptors. The conscious perception of pain in the brain, on the other hand, is a subjective sensation that varies greatly from individual to individual. The nociceptive nerve endings in the tongue and oral mucosa originate from the trigeminal nerve and the sensations are, therefore, often called trigeminal stimuli to distinguish them from taste. In the mouth, they serve mainly to prevent us from biting our tongue or burning our mouth with food that is too cold, too hot, or too acidic. The receptor molecules on the nociceptors are mainly ion channels that are opened or closed by stimulation and thus directly influence nerve activity (Roper 2014). The best studied channels on nociceptors are shown in Fig. 2.4. Since no signaling chain is built in between stimulation and excitation of the cell, the amplification factor, which is, for example, used in type II taste cells, is missing. The trigeminal nociceptors, therefore, react very quickly and can map a continuous spectrum of different stimulus intensities.

Temperature Sensation
TRP channels are cation channels which, when stimulated, mainly allow sodium ions to flow into the cell from outside. In doing so, the nerve cell is electrically excited. Some TRP channels are thermal sensors: They open and close depending on the ambient temperature. TRPA1 is responsible for extreme cold and TRPV1 for extreme heat. The fact that both channels are expressed simultaneously in over 70%

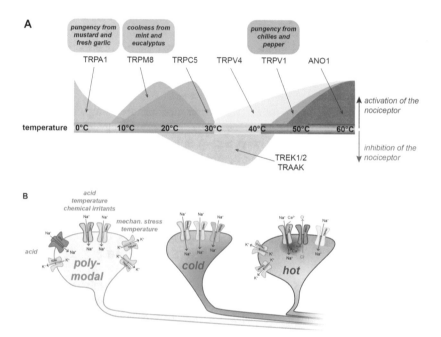

Fig. 2.4 Trigeminal stimuli in the oral cavity. **a** Thermoreceptors expressed on nociceptors with the temperature ranges and chemical stimuli that lead to the opening of the channel. **b** Three examples of different nociceptors with the respective receptor equipment (ANO1 = Anoctamin 1; ASIC = acid-sensing ion channel; TASK = acid-sensitive potassium channel; TREK, TRAAK = thermomechanical potassium channels; TRP = transient receptor potential channel). *Source* Own representation

of nerves explains our difficulty in distinguishing painful heat from cold. However, there are also nerve fibers that specialize in warmth and those that specialize in coolness. Warmth, that is, temperatures from 30 °C upward, is registered by TRPV4 and coolness, that is, temperatures below 30 °C, by TRPM8 and TRPC5. Temperature registration is moderated by the potassium channels TREK and TRAAK. Both open in a comfortable temperature window between 20 and 40 °C and allow potassium ions to flow out of the cell. This counteracts the electrical excitation and prevents our own body temperature from causing us pain. In the heat sensors, the signal is amplified by a chloride channel (Anoctamin 1). Chloride ions are negatively charged and when they flow out, the corresponding nerve cell is also excited. The chloride channels are temperature-controlled by themselves and open at similar temperatures as TRPV1. They are also located directly adjacent to TRPV1, which also allows some calcium ions to flow in. Calcium ions bind to anoctamine 1 from inside and additionally activate the channel. The heat sensors are, therefore, inhibited below 40 °C by TREK and TRAAK, but become quickly maximally active at temperatures above 43 °C and report a painful burn in the oral cavity. Temperature stimuli between 15 and 30 °C are not directly harmful to our body surfaces, but can contribute to (under)cooling the body. They must therefore also be registered and regulated according to the heat content of the body. This is what the cold receptors with TRPM8, TRPC5, and TRPA1 are for. They are not significantly inhibited by potassium channels, but they only trigger a slightly alarming feeling of coolness in the brain.

Mechanical Stress
In addition to temperature, nociceptors also register acid, other chemical irritants, and mechanical stress. Mechanical stress is caused, for example, not only by the bite on the tongue but also by a too high salt concentration. Such hyperosmolar solutions literally dry out the cells so that they shrink and shrivel. Such deformations of the cell membrane are also registered by the potassium channels TREK and TRAAK, which open or close depending on the direction of the deformation, thus making the respective nociceptors more difficult or easier to excite.

Acid Sensation
Acid is an excess of protons (H_3O^+ ions) or a low pH value. If the protons are only produced outside the cells, then it is acidic for the cells on the outside, but not on the inside. In this situation the two sodium channels TRPV1 and ASIC (acid-sensing ion channel) open and the potassium channel TASK closes. Such an acidification is mainly caused by strong, inorganic acids such as hydrochloric acid or phosphoric acid, as these cannot diffuse inward through the cell membrane. Weak, organic acids such as vinegar, citric acid, or CO_2 can diffuse into the cell interior and acidify it. If it becomes more acid inside the cell than outside, TRPA1 opens. Many nociceptors express TRPV1 and TPA1 simultaneously and are therefore unable to differentiate between strong and weak acids. However, with heat and cold sensors this differentiation is possible, so that carbonic acid (CO_2), acetic acid, or citric acid cause different sensations than, for example, phosphoric acid.

Chemical Stimuli/sharpness
Some chemicals in our environment can unspecifically change cell structures and thus act as poisons that we should avoid. These include organic solvents such as

alcohol, which change cell membranes, and molecules that attach themselves uncontrollably to our proteins or lipids, such as reactive oxygen species, heavy metals, or mustard oil. These stimuli are also recognized by the trigeminal nociceptors and their ion channels. Alcohol acts mainly via TRPV1, while TRPA1 seems to be responsible for the others (Bandell et al. 2007).

Some plants synthesize chemical molecules that activate the TRP channels even without heat or cold. This helps the plants not to be eaten by us or other animals. The best known pungent plants are some types of peppers, pepper, garlic, and mustard. Chilies and other hot peppers synthesize the molecule capsaicin, which activates TRPV1 and makes us believe they are unbearably hot. Piperine from peppercorns also acts via TRPV1. The isothiocyanates from mustard or allicin from fresh garlic, on the other hand, activates TRPA1 and thus a pungency that is more similar to that of freezing cold. Menthol, eucalyptol and many other essential oils bind and activate TRPM8 and cause a feeling of coolness.

Coke for hot food?

The answer depends on whether the food is too spicy or not spicy enough for you. Since typical cola drinks today are adjusted to a pH value of 2.5 using phosphoric acid instead of citric acid, the drink contains approx. 3mM free protons (H_3O^+), which acidify the outside but cannot penetrate the nerves to any significant extent. Cola thus additionally activates TRPV1 or at least makes it more sensitive. The pungency of the chilies is thus increased. Since the pungent molecule capsaicin is almost insoluble in water, it is not washed away from its receptor by cola. If you want to reduce the pungency, you should go for something fatty, for example, a piece of cheese or fatty yogurt.

It was certainly not the intention of these plants that we humans should enjoy the trigeminal stimulus in moderation. But for us, a certain pungency (in moderation) is welcome, as is the tingly feeling of effervescence or the slight bite of ethanol in drinks. Every baby and toddler will show you unequivocally that trigeminal stimuli are by nature negative at first. Only in later years of life do we learn to like some of these stimuli. The "why" can be speculated about quite a lot. Since spicy food is especially popular in warmer climates, eating it could help cool the body down. In fact, a TRPV1 agonist like capsaicin can lower the body's core temperature by increasing heat release through increased skin circulation and sweating. It is also often argued that pungency is toxic to bacteria and fungi, which means that food spoils more slowly in warm areas. However, there is no evidence of antimicrobial or fungicidal activity of capsaicin or piperine. However, chili pungency of a food causes it to pass through the gastrointestinal tract more quickly, so pathogenic germs in the food can cause less damage. Paul Rozin as a psychologist has another explanation for spicy food: He calls it "benign masochism," a special characteristic of humans that has not yet been observed in other animals. He says that we enjoy mastering negative emotions (spicy food or even a roller coaster ride) when they take place under safe and controllable conditions (Rozin and Schiller 1980). In fact, such activities release endorphins to a not inconsiderable extent and can even be addictive.

"After-Burner"

is a cheaky description of the burning sensation that occurs when going to the toilet after a very spicy meal. Although we have nociceptors with the corresponding TRP channels in the entire digestive tract from mouth to anus, we only become aware of the pain in the mouth and on the last centimeters before it leaves our body. An after-burner is only caused by those molecules that survive the gastrointestinal tract unchanged. These are mainly the capsaicin from the chilies and piperine from pepper. Mustard oil, allicin and also the essential oils are absorbed in the intestines and for the most part metabolized in the liver.

Astringent

The surface of the mouth seems to contract and dry out during the dubious consumption of an unripe banana, green tea that has been infused too hot or black tea that has been steeped too long, or wines that are strong in tannin. This feeling is called astringency and is also one of the trigeminal stimuli (Schöbel et al. 2014). In this process, gallic acid or polyphenols with gallic acid groups bind to a not yet identified receptor on trigeminal nerve endings. In contrast to the other trigeminal stimuli, ion channels do not appear to be directly involved in the binding of gallic acid, but rather G-protein-coupled receptors as in classical taste. Another important part of the astringent sensation may also be of a purely mechanical nature: The polyphenols bind to proteins from the saliva and precipitate them. This causes the saliva to lose its properties as a lubricant and the surfaces in the mouth feel rough and dried out.

Barrique

High quality red wines can be aged in barrels made of roasted oak. During this process, the oak releases astringent tannins and the roasting of the wood (toasting) produces the odorous substance vanillin. Together they make up the wine aroma "barrique." To avoid the expensive and time-consuming storage in wooden barrels, the red wines can also be stored in stainless steel barrels with toasted oak shavings.

Sparkling
Certain plant substances, such as hydroxy-α-sanshool from Szechuan pepper or Spilanthol from Jambú (Acmella oleracea), cause an anaesthetic pungency, which is often described with an electrifying sensation like small electric shocks on the tongue. There is evidence that the tingling molecules can directly activate TRPV1 and possibly TRPA1, but the specific electrifying effect seems to be caused primarily by their inhibition of potassium channels such as TASK in mechanically sensitive nerve endings without TRPV1 or TRPA1 (Albin and Simons 2010).

2.5   Taste Tricks From Nature, Science, and the Food Industry

In Nature
Plants use the sense of taste of animals to "talk" them into helping or to dissuade them from actions that are harmful to the plant. The incorporation of bitter toxins into plant parts that are not intended to be eaten by animals has already been reported in Sect. 2.2. However, many plants rely on animals to spread their seeds. For example, an apple often falls not far from the trunk of the parent tree, but cannot germinate there. The sweet flesh of the apple attracts animals, which often carry the delicious fruit away to eat. The apple seeds, however, are bitter and poisonous and are therefore usually not eaten. The chili plant has solved this problem in a different way: Its red flesh is not sweet and the seeds are not very well protected. The preferred distribution of the seeds is by birds. With a few exceptions, birds cannot taste sweet anyway, but they prefer red fruits with umami flavour. Birds often swallow the seeds of the chilies, but this is not a problem: Birds do not chew and do not have any significant stomach acid that would destroy the seeds. The seeds are excreted in a natural way (together with useful fertilizer) elsewhere. In order to protect itself from chewing animals with stomach acid, chili incorporates capsaicin, which tastes very hot for mammals, into the flesh of the fruit, which birds do not notice.

In Science
Among other things, the researchers are investigating how the sense of taste can be used diagnostically. In 2017, Ritzer et al. presented a chewing gum which, in the presence of inflammation in the oral cavity during chewing, releases the extremely bitter denatonium (see Fig. 2.3) from a tasteless precursor molecule. Without the need for an expensive measuring device the patient himself shows by his facial expression whether an inflammation is present. The molecule contained in chewing

gum can be modified in such a way that it is split either at every inflammation or only if a very specific pathogen is present. A test designed in this way would be rapid, specific, and can be carried out without any technical equipment.

In the Food Industry

For several decades now, the primary concern has no longer been to use processed foods to ensure that the growing world population is supplied with healthy and nutritious food. More and more it has to follow fashion trends such as "vegan," "natural," "functional," "allergen free," and "low calorie." Nevertheless, food should and must continue to taste good in order to survive in the market. Therefore, some companies are testing new flavors in high-throughput processes on cells outside of a living being that express human taste receptors (Riedel et al. 2017). Promising candidates from these pretests must then be toxicologically tested and finally confirmed (or rejected) by trained taste testers. Most of the substances tested are originally synthetically produced molecules. Only a few natural molecules, that is, molecules synthesized by living organisms, are available for such test series. Often different companies are involved in the identification, testing, mixing, and use of flavorings in food (effa 2018), which have agreed among themselves to keep the chemical composition confidential. Therefore it is hardly possible to find out which molecules are hidden behind the additive "flavor."

Sweeteners

In addition to the energy-supplying, "real" sugars such as sucrose from fruits or glucose, which is released in the mouth from starch, many other molecules taste sweet to us (see Fig. 2.5). These other sweet molecules were designed by plants or food chemists and, although they taste sweet, they provide us with very little or no energy. They are known as sweeteners.

The **artificial sweeteners** are mostly created by chance in the laboratory. Saccharin, for example, was created as an intermediate product during chemical experiments with coal tar in 1879. The chemist Constantin Fahlberg had forgotten to wash his hands before eating after his long day in the laboratory and discovered that his bread and water tasted excessively sweet. His fingers were quickly exposed as the cause of this phenomenon and in the following weeks the substance itself was isolated and characterized in the laboratory (Fahlberg 1886). Cyclamate was created in Michael Sveda's laboratory in 1937, when he actually wanted to synthesize a fever-reducing substance. And aspartame was discovered by James M. Schlatter in 1965 when he actually wanted to produce the peptide hormone gastrin.

Neither have plants developed sweeteners to sweeten our lives. Most **natural sweeteners** (Behrens et al. 2011) come from families of molecules that have a

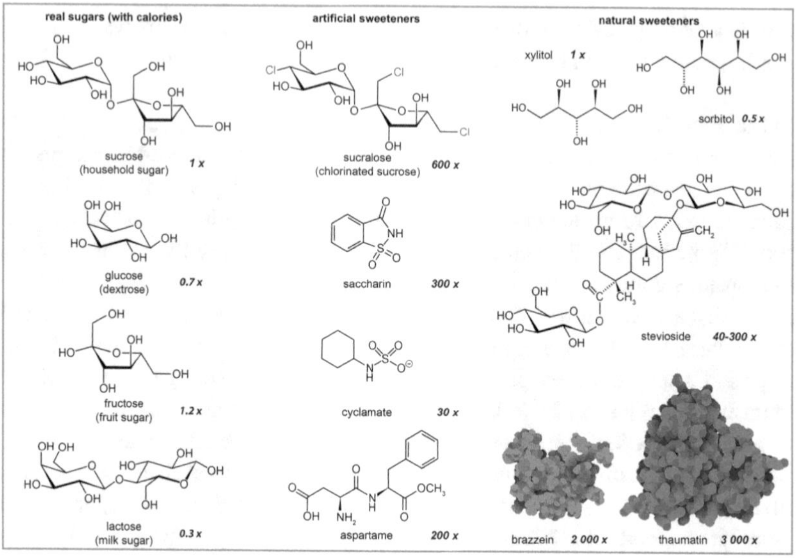

Fig. 2.5 Real sugars with calories and low-calorie/free sweeteners; the numerical value is the sweetening power of 1 g each compared to sucrose. *Source* Own representation

protective effect for plants: They are supposed to protect against predators, help against infections caused by viruses, bacteria or fungi, or reduce damage caused by UV light and bind water in the plant when drought conditions prevail or the soil is too salty. Plants usually produce a whole arsenal of different molecules and it seems to be a coincidence that some of them taste sweet to us humans. Natural sweeteners are therefore no healthier than artificial sweeteners. However, most sweeteners, whether artificial or natural, have a so much higher sweetening power than household sugar that just the minimum amount consumed virtually eliminates harmful side effects. The only exception are sugar alcohols (xylitol, sorbitol). They do not have a particularly high sweetening power and must, therefore, be consumed in quantities similar to those of sucrose. This means that they can also be used to bake cakes, but because they remain in the intestine and bind water so strongly, they often cause diarrhea. Terpene glycosides, such as the steviosides from the stevia plant or the mogrosides from the monk's fruit, are quite variable in their sweetening power depending on their composition and have a distinct bitter aftertaste. They are, therefore, not suitable for sweetening all food and drinks. Surprisingly, it was discovered that certain proteins from plants (thaumatin, brazzein) have a very strong and long-lasting sweet taste. One molecule of thaumatin, for example, is about 100,000 times sweeter than one molecule of sucrose. However, the protein surface does not look similar to that of the sugar molecules. The sweet proteins apparently do not bind to the same site as sugars and smaller sweeteners, but bind further down on our sweet receptor and clamp it in the active conformation for a longer period of time (Goodsell 2016). Another bizarre discovery is the taste-transforming proteins miraculin and neoculin. Miraculin does not taste at all at neutral pH, neoculin only tastes slightly sweet. Only when acid is added, both proteins activate the sweet receptor. But this amazing phenomenon is also an evolutionary coincidence, because it only works on the sweet receptor of some closely related primates, including humans.

Since the sugar lobby and the sweetener lobby have been fiercly fighting each other for over a century, there is a good many comparative studies on the effects of the two groups of sweet molecules in humans and animals. In the mouth, both sugar and sweetener molecules bind to the same sweet receptor and convey the same conscious sweet taste in our brain. Mice that had been genetically switched off the sweet receptor were consequently no longer able to taste sweeteners. Strangely enough, however, the loss of the sweet taste of "real" sugars did not completely disappear in these mice. A possible explanation: Taste cells also use sugar for their own energy supply. They can take up glucose from saliva via a passive transporter (GLUT) and a transporter driven by sodium ions (SGLT1) and extract ATP from it as intracellular energy currency. The formed ATP is not only an internal energy supplier but also the neurotransmitter that excites the associated nerves. In addition, ATP is able to close certain potassium channels in the membrane and thus activate the cell completely without the sweet receptor signalling pathway. The involvement of a sodium-dependent transporter in the sweet taste explains why common salt (sodium chloride) acts as a flavor enhancer for sugar and why a pinch of salt is an essential part of every sweet dish. There are, therefore, differences in the effects of sugars and sweeteners: Only calorie-containing sugars trigger the tiny but measurable increase in insulin, which is only mediated by taste sensory cells in the mouth (the so-called cephalic phase-insulin secretion) and also in the further course of digestion only sugars, but not the sweeteners, have an effect on the release of hormones in the body. And only the "real" sugars activate the dopaminergic reward system in the brain (Han et al. 2018). As a result, the concerns raised by the sugar industry that sweeteners could lead to hypoglycemia or even be addictive are refuted.

Umami Substance
Umami is also a very popular taste, which makes the non-sweet foods attractive. Meat, other animal foods, but also yeast, celery or tomatoes contain particularly

high concentrations of the umami molecules glutamate, inosine-, and guanosine-monophosphate. However, these must first be released by extensive chewing and salivation in the mouth. A raw piece of meat or fish, therefore, tastes bland at first. Roasting, cooking, pickling, or sun drying can already breakdown a few large molecules and thus make the pretreated food more palatable. In order to achieve an umami taste in food that actually has hardly any umami ingredients, without any complex pretreatment, you can use "seasoning": This is the mixture of the three umami taste molecules in free form. Other umami substances have been found in heated or otherwise processed foods with a strong spicy taste. These are mainly produced by the combination of sugars with amino acids in the so-called Maillard reaction (Behrens et al. 2011). With the alapyridine from glucose and alanine a real flavor enhancer was found, a molecule that has no taste of its own, but which enhances the umami (and sometimes also sweet) taste of other molecules. A particularly strong and long-lasting umami taste is also produced by molecules that are formed from the nucleotide guanosine monophosphate and acetic or lactic acid when heated. They were found in dried and fermented tuna meat from Japanese cuisine (see Fig. 2.6).

The Chinese Restaurant Syndrome

gives the umami taste a bad name. In the late 1960s, there were several case reports in the medical literature about acute complaints with reddening of the skin, dry mouth, and palpitations after visiting Chinese restaurants. In 1969, Schaumburg et al. published detailed studies in humans and found it to be proven that the added sodium glutamate was responsible for the symptoms. Since we take in about 10g of sodium glutamate every day with our normal diet without symptoms, the hypothesis was not convincing from the beginning. Although it is not known what causes the symptoms from the first case reports in the 1960s, double-blind studies have since disproved the causal relationship with glutamate. In some people who claim to suffer from the syndrome themselves, it seems to be mainly a nocebo effect. The mere fear that glutamate is present in food as a flavor enhancer leads to supposedly glutamate-specific complaints. Although, according to scientific data, sodium glutamate is to be classified as completely harmless to health, the umami substances cannot get rid of their bad reputation. In order to be able to market clear broth and seasoning after all, many dishes were soon offered with yeast extract, which naturally contains glutamate and nucleotides. A short time later, yeast extract was also unmasked as "the new flavor enhancer" and had to be replaced. Without yeast, glutamate, and nucleotides, only celery powder remains as seasoning. But even the previously inconspicuous celery tuber would soon be caught in the crossfire, as it was praised in a ZEIT article (Fassbender 2018) as "natural glutamate" and with the words "more umami is not possible." In addition, celery must be labeled as an allergen. So back to pure chemicals after all?

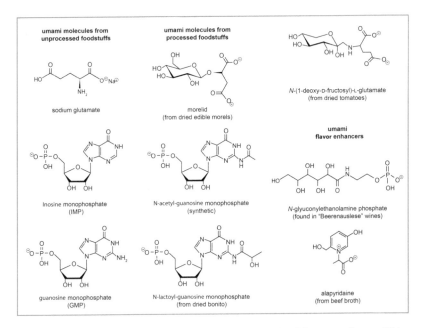

Fig. 2.6 Umami substances; are usually chemically synthesized for use as flavor additives for foods and are not isolated from the foods mentioned. *Source* Own representation

Bitter Blocker

The bitter taste of, for example, drugs or certain foods or beverages can be at least partially suppressed in various ways (Bennett et al. 2012):

A) Binding of the bitter substance to another molecule so that it is no longer available for binding to the bitter receptor;

B) Direct blocking of the binding site for the bitter molecule at the receptor by another molecule that does not activate the receptor (antagonist);

C) Simultaneous stimulation of other tastes, which, by interfering in the taste bud or even in the brain, make the bitter taste less obtrusive.

In variant A, it depends on the properties of the bitter molecule whether the trick works. Some bitter substances such as quinine are more soluble in fat than in water and can thus be prevented from entering the aqueous saliva and thus the receptor by adding fat. This trick does not work with bitter substances that are highly soluble in water, such as the caffeine in coffee. The use of cyclodextrins is even more specific for the individual bitter substance. These are ring-shaped molecules made up of sugar units that, depending on the size of the ring, provide space for certain bitter substances in their interior. The bitter substances are thus hidden in a sugar ring.

In variant B, the receptor molecules themselves are the target. At least 13 different molecules are known to block different bitter receptors (Jaggupilli et al. 2016). Most of them block only one of the 25 bitter receptors. One of the molecules can block ten different receptors, but "the" one bitter blocker for all 25 will most likely never be found.

Combined sweetness against the bitter aftertaste

In addition to their enormous sweetening power, the artificial sweeteners saccharin and cyclamate unfortunately also have a bitter aftertaste, as they also activate individual bitter receptors. However, a mixture of saccharin and cyclamate is considerably less bitter, which is why the two sweeteners have been predominantly offered in combination for many decades. Behrens et al. (2017) have now been able to show that both molecules are also active as bitter blockers. Cyclamate blocks the bitter receptors that are stimulated by saccharin and vice versa, saccharin inhibits the bitter receptor activated by cyclamate.

Variant C uses our own way of perception. As already mentioned, different taste cells influence each other already within a taste bud and also in the brain taste information is evaluated contextually. With regard to the unloved flavors, acidic and bitter this means: The more positive taste the food has to offer at the same time, that is, sweet, umami, and possibly also salty, the less unpleasant the acidic and bitter taste is for us. A spoonful of sugar can therefore make not only coffee but also a broccoli dish taste much more pleasant. And who has not experienced it yet: Orange juice tastes only acidic and bitter immediately after brushing one's teeth. The detergents in the toothpaste have temporarily disabled the sweet receptors.

With bitter blockers there is no resistance from traditional food companies as there is with the sugar industry and sweeteners. Nevertheless, there are also concerns here as to whether switching off the bitter taste really is ultimately positive for us humans (Molteri 2016). Because it is certainly sensible to limit the consumption of bitter substances (=poisons!) as much as possible. Bitterness is more than just a bad taste that no longer fits into the modern lifestyle. Bitter molecules also influence digestion, metabolism, and immune defence (see Chap. 3).

Fat Substitutes

Fat is our most calorie dense food component and in times of slimming mania it has fallen into disrepute. Nevertheless, we simply like fried fries better than boiled potatoes. Would it not be great to have substitutes that mimic the taste and mouthfeel of fat, but not the calories? So far, only substances that can simulate the mouthfeel but not the actual taste of fat are in the market. Sucrose polyesters (fatty acids are attached to sucrose instead of glycerol, for comparison see Fig. 2.2) such as olestra, which is approved in the USA, have the advantage that they can also be used for frying and baking, whereas fat substitutes based on carbohydrates or protein are only suitable for cold dishes. The latter include the EU-approved fat substitutes carrageenan from red algae, simplesse from whey protein, maltrin and Paselli SA2 from starch and inulin, a vegetable storage carbohydrate. Whether it is the lack of fat taste in the mouth or the lack of calories, laboratory animals and humans unconsciously register the lack of fat and react with less feeling of satiety after the fat-reduced meals. Gourmets also have not yet been convinced of the equivalent taste quality of low-fat variants of a food (Starz, p. 2017).

Kokumi Substances
A positive mouthfeel and lasting richness of taste is mediated by ligands of the calcium sensitive receptor (CaSR). However, since calcium itself has a bitter aftertaste and is only soluble to a very limited extent (lime as calcium carbonate or lime soaps in combination with fatty acids), the food industry is looking for other molecules that activate the CaSR. Even before the kokumi taste impression was associated with a calcium receptor, Ueda et al. discovered in 1990 kokumi flavors in garlic extracts that were tasteless in themselves. However, when this extract was added to glutamate and inosine monophosphate, the test subjects were able to perceive a distinct garlic flavor with the typical mouthfeel of kokumi. In this first experiment, sulfur-containing molecules were published as the decisive kokumi substances. Among them was glutathione, a peptide consisting of the three amino acids γ-glutamate-cysteine-glycine. However, these also always had a sulfurous odor component, so that at the beginning of the research this odor was associated with kokumi. The following intensive search for molecules with kokumi properties showed that it was not the sulfur but the γ-glutamate in the peptide that was the decisive factor. Currently, the peptide consisting of γ-glutamate, valine, and glycine is the strongest known kokumi substance (Amino et al. 2016). New foods containing this peptide are being discovered all the time. Currently fish and soy sauce, yeast extract, shrimp paste, and cheese are at the top of the list.

Tasting in Unusual Places

3

Summary

Since the first indications of taste receptors and signaling components outside the mouth, "tasting" cells have been identified in almost every organ studied. These usually taste the same molecules as the classical taste cells in the mouth and give the organ detailed information about the current local nutritional situation. Wherever bitter receptors are involved, it is usually—as in the mouth—a warning of toxins or even pathogens. However, scepticism is appropriate if the classic sweet and umami receptors from T1Rs are held responsible for effects within our body. These receptor dimers are quite insensitive to sugar and amino acids and are only activated at higher concentrations than those that occur within the body.

The discovery and clarification of extraoral taste opens up completely new possibilities for the development of specific drugs. For example, bitter substances that relax smooth muscle cells in the bronchi may soon be found in asthma sprays (Shaik et al. 2016). However, to derive nutritional recommendations from such very specific, local effects of taste molecules on individual receptors in individual organs would be shortsighted. If the diet is changed, this affects all organs in the most complex way—we cannot predict most of the effects and they may be potentially harmful to the organism as a whole.

3.1 Tasting Beyond the Mouth in the Stomach and Intestines

There are two types of cells in the stomach and especially in the intestine, which contain taste receptors and the necessary elements of the signaling cascade: The solitary chemosensitive cells (also called tuft-cells) and the enteroendocrine cells,

© Springer Fachmedien Wiesbaden GmbH, part of Springer Nature 2021
P. Schling, *The Sense of Taste,* essentials,
https://doi.org/10.1007/978-3-658-32233-5_3

which can release hormones into the bloodstream. In the digestive tract, the sense of taste, that is, specialized cells with taste receptors, has two important tasks: First, they regulate the digestive processes and the feeling of satiety depending on the food that arrives, and second, they are important sensors and control centers for coexistence with all the viruses, bacteria, other unicellular organisms, and worms in the lower sections of the intestine.

Coordination of the Digestion and the Degree of Filling of the Stomach and Intestines
Our sense of taste on the tongue decides above all whether we swallow or spit and only the information from the mouth lets us become aware of what we are eating. However, when we have swallowed something, the stomach and intestines do not simply accept it, but continue to control and regulate it all the way. The most important parameters are the speed of gastric emptying (and in which direction), the amount of digestive secretion from the pancreas and gall bladder, the speed and strength of intestinal movement, the equipment of the intestinal cells with transporters, and how quickly the ingested substances are transported out through the large intestine. Bitter and sharp molecules accelerate the intestinal passage, while fatty and umami tend to slow down to ensure complete absorption. Glucose leads to an upregulation of the corresponding sugar transporters in the small intestine so that none of the sweet molecules are lost to us. In the case of the satiety hormones, the effects for bitter, pungent, sweet, fatty, and umami often go in a similar direction, because we should stop eating both when the food is toxic or painful as well as when we have absorbed enough energy and building blocks (Steensels and Depoortere 2018).

Worm Alarm
The solitary chemosensitive cells can apparently "taste" harmless protozoa (small unicellular organisms) and unwanted worms. They have bitter and umami receptors, but which molecules bind to which receptors and betray a worm is still unknown (Lu et al. 2017). While they are only moderately active with symbionts to keep their numbers within reasonable limits, they sound the alarm in the case of dangerous parasites. For this purpose, they release an increased amount of a messenger substance, interleukin 25 (IL25). IL25 introduces a special class of immune cells, the "innate lymphoid cells type 2" (ILC2), which now release another interleukin, IL13. IL13 in turn leads to the proliferation and activation of goblet cells (mucus-producing cells) and also of solitary chemosensitive cells (positive feedback!). In the resting state (without worm), the epithelium in the small intestine consists mainly of the resorptive epithelial cells, which busily take up food molecules and release them

into the blood. Goblet cells make up 4–10% of the cells and the solitary chemosensitive cells less than 1%. The IL13 from the ILC2 doubles the number of goblet cells and increases the number of solitary chemosensitive cells tenfold. The goblet cells not only secrete more mucus but now also produce special mucus molecules that make the mucus tougher and more inedible for the worms. This results in the worms being discharged with the stool, typically within 10–14 days.

Bitter as an Aperitif or the Herb Bitter Afterward?
Among other things, an aperitif should stimulate the appetite. Among the classic aperitifs, there are also bitter variants with Aperol and Campari. The short increase in the hunger hormone ghrelin, which endocrine cells secrete from the stomach in response to bitter substances, probably has an effect here. Ultimately, however, the bitter substances inhibit gastric emptying and promote a feeling of satiety. A bitter alcoholic drink after a sumptuous meal can, therefore, also prolong the feeling of fullness. Once in the intestine, however, it promotes the digestion of fatty foods in particular, as it increases the supply of bile and stimulates intestinal movement. It all depends on the right timing.

3.2   How Sweet Can It Be? Pancreas and Brain

All our cells like sugar and actually we cannot get too much of it. Therefore, the intestinal absorption capacity is up-regulated when a lot of sugar is registered there. The pancreas regulates which organs are allowed to feed on the sugar in the blood: If the glucose concentration is high (more than 5–6 mM), the β cells of the pancreas release their hormone insulin, then the heart, skeletal muscle, and fatty tissue are also allowed to absorb and store the glucose molecules. When the blood sugar level is low (less than 5 mM) without insulin, the sugar is left for the others, including above all the brain and the red blood cells, which are not able to survive without glucose.

The β cells of the **pancreas,** therefore, have a sophisticated measurement system for the blood sugar level that is not yet understood in detail. Glucose can bind to a sweet receptor on the cell surface or can be absorbed into the cell to be metabolized. Both are important for an adapted insulin release. The sweet receptor in β cells does not consist of the classical subunits T1R2 and T1R3 as in the mouth (Kojima et al. 2017). This receptor mainly binds to sucrose, the classic household sugar. The blood sugar glucose, however, activates it only at concentrations around 50 mM. Even severely diabetic patients do not have such a high blood sugar level. Besides T1R3, the β cells of the pancreas express the calcium-sensitive receptor (CaSR). In addition to calcium ions, this receptor can also bind to glucose, even at much lower concentrations of about 3 mM. Investigations suggest that, in addition to pairs of similar receptor proteins T1R3/T1R3 (too weak glucose binding) and CaSR/CaSR (too strong glucose binding), mixed pairs of T1R3/CaSR are formed in the cell membrane of the β cells. These pairs are activated at glucose concentrations around 8 mM. This enables the pancreas to register very precisely when and how much insulin it should release. The sweet receptor consisting of T1R3 and CaSR binds artificial sweeteners such as sucralose much more weakly than the classical sweet receptor consisting of T1R2 and T1R3. Since only about 5% of the tiny amounts of artificial sweeteners used to sweeten food and beverages are absorbed at all in the intestine, they do not reach the necessary concentrations in the blood to cause the pancreas to release insulin.

The information about how much sugar we still need to satisfy all the cells in the body is ultimately collected in the brain. There is a collection of nerve cells called the **hypothalamus**. The hypothalamus contains the set screws for important control circuits, including hunger and satiety for food intake. For this purpose, certain neurons in the hypothalamus also measure blood sugar. If this is too low, our motivation to consume more carbohydrates increases: We get hungry. How these neurons measure glucose concentration is still poorly understood. They are equipped with sensors similar to those of the β cells of the pancreas and probably react primarily to glucose uptake and metabolism. On their surface, however, they also carry the classic sweet receptor consisting of T1R2 and T1R3, which, however, is unlikely to react to normal blood sugar fluctuations and whose function is therefore still unclear. Incidentally, the hypothalamus does not regulate the two sensations of hunger and satiety equally strictly: If the blood sugar level drops, it triggers a strong feeling of hunger, if the blood sugar level rises, hunger decreases. That is why we can think of something other than food and not actively seek it. But if something tasty, sweet, and fatty is unexpectedly in front of us, we still reach for it. A well-filled calorie storage for worse times cannot hurt from the point of view of our hypothalamus.

3.3   The Taste of Bacteria—Nose and Lungs Are Alert

With every breath we take, our airways come into contact with pathogens that would love to colonize the moist, warm, and nutrient-rich mucous membranes. The first line of defense against this is the uppermost cell layer, the epithelium, with its innate, unspecific immune system. However, since there is always collateral damage from every fight, not every single bacterium should immediately cause a fulminant sinusitis or pneumonia. So our nose and also the lungs can "taste" how many bacteria are currently living on the surface or in the mucus (Lee and Cohen 2014) and intensify the defense reactions accordingly if necessary.

The most common cell type lining the surface of the nose and lungs is the cilia cell. The goblet cells are located between the cilia cells. They secrete a tough, sticky mucus (gel layer) that floats on an aqueous, very liquid sol layer. On their surface, the ciliary cells carry thin, actively mobile cell protrusions that protrude into the gel layer in full length. Through a coordinated movement toward the throat, they continuously move the mucus upward. After such a cilia impact, a cilium can then be bent and withdrawn through the sol layer with almost no resistance. Most inhaled pathogens and dirt particles remain stuck in the mucus and are then transported by the cilia toward the throat to be swallowed, coughed up, or sneezed out. This process is called "mucociliary clearance." In addition to this mechanical defense, the cilia cells also possess at least two potent chemical warfare agents: The radical nitric oxide (NO) and the antimicrobial peptides (AMPs). Both substances kill bacteria and other pathogens directly, thus preventing them from multiplying too quickly during transport.

Epithelia can "taste" how many bacteria are currently living in the mucus (Fig. 3.1). This determines how fast the cilia beat and how much NO and AMPs have to be secreted. Cilia cells have bitter receptors on their protuberances for this purpose; in humans these are mainly the bitter receptors with the numbers 4, 38, 43, and 46, number 38 (T2R38) being particularly important for the cilia cells in the paranasal sinuses. Without T2R38 these would be virtually defenseless. But why do bacteria taste bitter? They produce a molecule called n-**A**cyl-**H**omoserine-Lactone (AHL, see Fig. 3.2). Bacteria use AHL to determine their own population density. This effect is called quorum sensing and regulates the division of work between bacteria of a type, such as the formation of biofilms, which only makes sense above a certain cell density. The fact that AHL bind to T2Rs and thus also tastes bitter is certainly not intended by the bacteria, because in this way they also give our surface cells information about how many they are and whether it is worthwhile to increase the defense reactions.

When AHL molecules bind to the bitter receptors of cilia cells, this leads to an increase in free calcium ions inside the cell via the signaling pathways described above (see Sect. 2.1). On the one hand, this calcium accelerates the cilia beat and on the other hand activates an enzyme that accelerates the formation of NO to kill the bacteria.

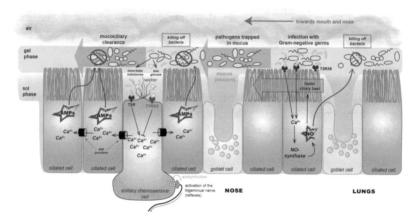

Fig. 3.1 Respiratory defense mechanisms and regulation by bitter and sweet receptors; abbreviations: AMPs, antimicrobial peptides; AHL, n-acyl homoserine lactone; Ca^{2+}, calcium ions; NO, nitric oxide; T1R2/3, sweet receptor; T2R; bitter receptor. *Source* Own representation

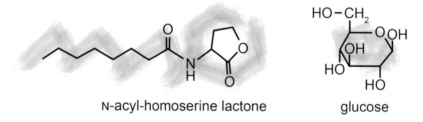

N-acyl-homoserine lactone   glucose

Fig. 3.2 "Taste" molecules in nasal and lung mucus that provide information about the density of bacteria. *Source* Own representation

Tasters and non-tasters

Students usually learn about the **bitter receptor number 38 (T2R38)** when Mendel's rules of inheritance are explained. Everyone is given test strips containing the chemical phenylthiocarbamide (PTC) and is allowed to test whether the chemical tastes bitter. The same then at home with the parents. There are only two variants of the gene coding for T2R38 in the fair-skinned European population: A functional and a defective one. Homozygous (two intact copies of the gene) and heterozygous (one intact and one defective copy) carriers of the functional variant can taste PTC, the others cannot. Non-tasters account for about 30% of the population.

Non-tasters, however, not only find the paper strips boring, they also lack one of the possibilities to measure the density of bacteria in the nose and lungs and adjust the cilia beat accordingly. Therefore, they suffer more frequently and severely from chronic **sinusitis**. And new drugs based on bitter substances, which are supposed to improve the removal of mucus from the nose and bronchi, will hardly help them (Shaik et al. 2016).

In addition to the direct stimulation of the ciliary cells by bitter substances, real sensory cells are also found in the epithelia of the respiratory tract. They do not have their own defense mechanisms, but they transmit the signal to neighboring cells and nerves. Under normal conditions they make up about 1% of the cells in the respiratory epithelium. Because they are so lonely, they are called solitary chemosensitive cells. They look very similar to type 2 sensory cells in the taste buds. Above all, they also have the distinctive tuft of fine cell protuberances (microvilli) that protrude outward and carry the taste receptors. This is why they are also called brush cells or tuft cells. They not only have different bitter receptors to taste the AHLs but also the classic sweet receptor, which measures the glucose concentration. Glucose thereby dampens the reaction of the solitary chemosensitive cell to bitter substances. How exactly this happens is not yet understood. Since bacteria consume the glucose concentration in the fluids on the nose and lungs, they might be able to increase the sensitivity of the solitary chemosensitive cells to AHL and hence the transmission of signals to neighboring cells and the call for a defense reaction.

Also in solitary chemosensitive cells, the binding of bitter substances to T2R leads to an intracellular increase in calcium ions. These ions enter neighboring ciliary cells via channels between cells, the so-called "gap junctions," and stimulate the release of antimicrobial peptides (AMPs). At the same time, they also electrically excite the sensor cell and activate (at least in mice) trigeminal nerves via acetylcholine release. Indirectly, the bacteria thus tickle the nose and are ejected in a high arc by the sneezing reflex triggered.

Below the mucus layer and the epithelial cells, there is a layer of smooth muscle cells in the lower airways of the lungs. They regulate the ventilation of the deeper lung areas through their contraction or relaxation. They also carry T2R receptors and can, therefore, also react to an increase in bacterial density. Although these cells also respond to stimulation with bitter substances with an intracellular increase in calcium, a situation that has always been associated with contraction, they relax when the signal comes from the bitter receptor. How exactly calcium ions can have one effect and the other is an exciting field of research. It is probably the exact localization (where) and the exact time (when) of the calcium impulse that makes the difference. Ultimately, here—in contrast to all other cells—the bitter signal cascade counteracts electrical excitation of the cell and leads to its relaxation. Bitter substances from bacteria thus open up the respiratory tract so that the infected mucus can also be efficiently transported away by the cilia.

Asthma

is a disease in which these smooth muscle cells contract too much, multiply excessively, and make it difficult for mucus and pathogens to be adequately transported from the lungs. The contraction is caused by a variety of different local factors. In an asthma attack, it is hardly possible to inhibit all of them simultaneously, which is why mainly direct bronchodilators are used, mostly agonists for the β_2-adrenergic receptor. Unfortunately, however, the β_2 receptors are downregulated in chronic inflammation, which often occurs in asthmatics. The effect of the drug is, therefore, often limited. Bitter substances could help here. The bitter receptors are not or only slightly influenced by inflammation, and bitter substances have a significantly better effect in an asthma attack than the β_2 agonist, at least in the mouse model (Shaik et al. 2016).

The **urethra** is the connection of the bladder to the outside and there is a risk of invading bacteria. Here, sensory cells are very similar to those in the nose and lungs. They are called urethra crested cells (urethral brush cells = UBS). Like their relatives from the respiratory tract, they can "taste" bitter, but umami instead of sweet. Both stimuli pose a danger to the urinary tract: Many bacterial products are bitter and glutamate (umami) promotes bacterial growth in urine (Kandel et al. 2018). If the sensory cells are excited by bitter substances or glutamate, they release the messenger acetylcholine and stimulate connected sensory nerves. This triggers a reflex urge to urinate in order to flush the unloved fellow occupants out of the urinary tract. These cells are, therefore, the reason why a bacterial cystitis is associated with the constant feeling of having to urinate.

3.4   Bitter Medicine—Good or Bad?

You should not lose sight of the basic principle bitter = poisonous, because it also applies to medication, of course.

Most drugs taste bitter and are also poisonous for a healthy person. However, the toxic effect can have advantages in case of illness. The best example is malaria, which still claims hundreds of thousands of lives every year. Malaria parasites hide from our immune system in the red blood cells during one of their developmental stages. There they feed on the red blood pigment, hemoglobin.

While they eat the globin, the poisonous hem remains. Two bitter substances, quinine and chloroquine, prevent the malarial parasites from disposing of the hem, so that the parasite dies from it. Nevertheless, quinine and a little less chloroquine is quite toxic for the patient. Besides the typical effects of bitter substances (gastrointestinal complaints), quinine also inhibits relative unspecifically, various cation channels and thus the excitability of muscle and nerve cells. However, if quinine helps to survive the malaria infection, the side effects are accepted. The benefit:risk ratio is, therefore, positive for malaria.

PROP against hyperthyroidism

Since 2015 it has been known (Clark et al. 2015) that thyrocytes—the functionally most important cells of the thyroid gland—can also "taste" bitter substances. They carry several T2R, including T2R38, on their cell surface and react to the corresponding ligands with a reduced iodide transport in the follicles, in which the thyroid hormones are produced. Preliminary results of genetic studies suggest that individuals who carry nonfunctional receptor variants in their thyroid gland are more prone to hyperthyroidism. The biological sense of such an inhibition of thyroid function by bitter substances can so far only be speculated about. Conceivable are a reduced appetite, a slower absorption of food from the intestine, and a reduced metabolism in the case of a bitter substance-rich diet in order to reduce the risk of poisoning. Interestingly, two ligands of T2R38, PROP, and thiamazole, are used as drugs in hyperthyroidism. Besides binding to T2R38, they also inhibit the enzyme that incorporates iodine into thyroid hormones.

However, bitter substances are not healthy in principle, as is often claimed. They do not protect against cancer or atherosclerosis, nor do they prevent diabetes or heart attacks. Healthy people should not take medication or other poisons, because without the disease the benefit:risk ratio is always negative. And even if drugs have a curative effect on the disease, they will not help if the person as a whole dies in the process. So if you read that one or the other bitter substance kills cancer cells, then this is certainly true. Amygdalin from apricot kernels, for example, is a very effective poison that releases cyanide, the anion of prussic acid, in the intestines or cellular metabolism and thereby stops cell respiration in particular. Cancer cells also die from this—but not only these. There is a clearly negative benefit:risk ratio for lack of clinical benefit and high risk potential. In other words, amygdalin not only fights the tumor, but the whole person. It leads to severe poisoning and even death (Hübner 2015). So if the pathogens or cancer cells suffer more from the poison than the rest of the body when dosed correctly, then we have a drug, otherwise the bitter substance is and remains simply a poison.

It is, therefore, not possible to prevent drugs from tasting bitter. However, this has a considerable negative influence on the willingness of patients to take these drugs. The pharmaceutical industry has long been working on ways to mask the bitter taste (Chauhan 2017). Up to now, the easiest way to do this has been to add sweeteners or sugar, because chemical modifications of the bitter active ingredient or binding to an insoluble matrix inevitably change its effectiveness and require expensive new clinical test series. A bitter "taste" in the intestine can also reduce the effectiveness of a drug, as the intestine activates defense mechanisms to remove the toxin from the body as quickly as possible. Jeon et al. published an interesting discovery in 2008: In mice, the sensitivity of the intestine to bitter substances depends, among other things, on the cholesterol content of the food. Mice that take up low cholesterol with food upregulate their bitter receptors. The researchers speculate that the lack of cholesterol, which is only found in animal food, might indicate the potential danger from plant toxins: Low cholesterol = high plant food = many toxins. If the data can be transferred to humans, this could mean that the effect of drugs also depends on diet. Vegans or people treated with cholesterol-lowering drugs might need a higher dosage to achieve the same effect of a drug.

What You Learned From This *essential*

- Once we have decided to put something in our mouths, it is the taste, not the aroma, that determines whether and how much of it we actually eat.
- Although the sense of taste has been researched since the end of the nineteenth century, many questions are still unanswered. In particular, acidic and salty tastes are still poorly understood in molecular and cellular terms.
- We can only taste small molecules, the fragments of macronutrients. These are naturally released in the mouth through prolonged chewing and salivation, but they can also be present in the food before this happens by processing the food, such as heating or drying it in the sun or adding the taste molecules directly to the food.
- The human body has at least six tastes: **Sweet, fatty,** and **umami** indicate the three important macronutrients carbohydrates, fats, and proteins; **bitter** warns us of toxins; **salty indicates** the common salt content of food and is perceived positively when the body needs it; **acidic** as a taste is not only responsive to fruit acids but also to CO_2 and the water content.
- **Kokumi** as a positive calcium taste is not perceived on its own, but enhances the positive taste sensations sweet and umami.
- The sense of taste is modulated significantly by **trigeminal stimuli.** These include temperature sensation, painful acid sensation, and mechanical deformation, which are mediated via ion channels on free nerve endings. Certain chemical molecules can open the same ion channels and thus "taste" sharp, cool, astringent, and also electrically tingling.
- Food chemists are looking for molecules that make "healthy" food more attractive in terms of taste. The aim is to enhance positive tastes and block or mask negative ones, especially bitter. With the exception of sweeteners and a few

© Springer Fachmedien Wiesbaden GmbH, part of Springer Nature 2021
P. Schling, *The Sense of Taste,* essentials,
https://doi.org/10.1007/978-3-658-32233-5

bitter blockers, no groundbreaking success has yet been achieved in this area. This shows how difficult it is to deceive our taste.

- The stomach and intestines use the sense of taste to optimally adapt the digestion to the food taken in and to trigger a feeling of satiety. In the body, taste receptors in individual organs can help to distribute the food molecules in an appropriate way. Body surfaces that come into contact with potentially harmful germs have transformed the bitter taste as a sensor for the innate immune system. This allows bacteria to be drained from the nose in time and worms to be expelled from the large intestine.

Glossary

Adenosine triphosphate (ATP) molecule, which contains a lot of chemical energy and is used within cells as an energy currency. Outside of cells it serves as a messenger

Anion chemical particle with total negative charge

Enzyme Protein that can accelerate a specific chemical reaction

Epithelium Cell layers covering the body surfaces

Expression Use the genetic information on a protein to produce the protein and transport it to the right place in the cell

Cation chemical particle with positive total charge

Ligand molecule that can specifically bind to and activate a receptor

mM Abbreviation for millimolar = millimol/liter, a common unit of concentration

Molecule chemical substance consisting of several atoms which are connected with each other

Neurotransmitter Molecule secreted by activated sensory or nerve cells

Protein large polymer composed of single amino acid molecules

Receptor a protein or protein complex that can transmit external stimuli to an intracellular signalling chain

Stimulus/stimulate Stimulus/excite, activate

Trigeminal of the trigeminal nerve, a nerve cord that transmits information about temperature or pain from the tongue and other regions of the face to the brainstem (5th cranial nerve, triplet nerve)

Vesicles small membrane-enveloped "bubbles" in cells, which are mainly used to transport molecules through the cell and out of the cell

© Springer Fachmedien Wiesbaden GmbH, part of Springer Nature 2021
P. Schling, *The Sense of Taste,* essentials,
https://doi.org/10.1007/978-3-658-32233-5

Literature

Albin KC, Simons CT (2010) Psychophysical evaluation of a sanshool derivative (alkylamide) and the elucidation of mechanisms subserving tingle. PLoS ONE 5(3):e9520

Amino Y, Nakazawa M, Kaneko M, Miyaki T, Miyamura N, Maruyama Y, Etoa Y (2016) Structure–CasR–activity relation of kokumi γ-glutamyl peptides. Chem Pharm Bull 64:1181–1189

Baldwin MW, Toda Y, Nakagita T, O'Connell TMJ, Klasing KC, Misaka T, Edwards SV, Liberles SD (2014) Evolution of sweet taste perception in hummingbirds by transformation of the ancestral umami receptor. Science 345(6199):929–933

Bandell M, Macpherson LJ, Patapoutain A (2007) From chills to chilis: mechanisms for thermoregulation and chemesthesis via thermoTRPs. Curr Opin Neurobiol 17:490–497

Behrens M, Meyerhof W, Hellfritsch C, Hofmann T (2011) Sweet and umami taste: natural products, their chemosensory targets, and beyond. Angew Chem Int Ed 50:2220–2242

Behrens M, Blank K, Meyerhof W (2017) Blends of non-caloric sweeteners saccharin and cyclamate show reduced off-taste due to TAS2R bitter receptor inhibition. Cell Chem Biol 24:1199–1204

Bennett SM, Zhou l, Hayes JE (2012) Using milk fat to reduce the irritation and bitter taste of ibuprofen. Chemosens Percept 5(3–4):231–236

Boesvelt S, de Graaf K (2017) The differential role of smell and taste for eating behavior. Perception 46(3–4):307–319

Bouchard B, Lisney TJ, Campagna S, Célériera A (2017) Do bottlenose dolphins display behavioural response to fish taste? Appl Anim Behav Sci 194:120–126

Chaudhari N, Roper SD (2010) The cell biology of taste. J Cell Biol 190(3):285–296

Chauhan R (2017) Taste masking: a unique approach for bitter drugs. J Stem Cell Bio Transplant 1(2):12

Clark AA, Dotson CD, Elson AET, Voigt A, Boehm U, Meyerhof W, Steinle NI, Munger S (2015) TAS2R bitter taste receptors regulate thyroid function. FASEB J 29:164–172

DKFZ (2016) Magenkrebs: Risikofaktoren und Auslöser. Krebsinformationsdienst, Deutsches Krebsforschungszentrum. www.krebsinformationsdienst.de/tumorarten/magenkrebs/risikofaktoren.php. Zugegriffen: 23. Okt. 2018

effa (2018) Discover the world of flavourings. European Flavour Association. http://effa.eu/flavourings/world-of-flavourings. Zugegriffen: 17. Okt. 2018

Fahlberg C (1886) The inventor of saccharine. Am Sci 55(3):36

Faßbender W (2018) Sellerie – Tolle Knolle. ZEIT-online. https://www.zeit.de/zeit-magazin/essen-trinken/2018-05/sellerie-image-wandel-rezepte-gastronomie/komplettansicht. Zugegriffen: 16. Okt. 2018

Goldstein E (1954) Vergiftung durch gegrünten Dosenspinat. Fette, Seifen, Anstrichmittel 56(2):109–110

Goodsell D (2016) Monellin and other supersweet proteins trick our taste receptors. RCSB PDP-101 molecule of the month. https://doi.org/10.2210/rcsb_pdb/mom_2016_7. Zugegriffen: 12. Okt. 2018

Han P, Bagenna B, Fu M (2018) The sweet taste signalling pathways in the oral cavity and the gastrointestinal tract affect human appetite and food intake: a review. J Food Sci Nut Int. https://doi.org/10.1080/09637486.2018.1492522

Hübner J (2015) Stellungnahme der Arbeitsgemeinschaft Prävention und Integrative Onkologie (PRIO) in der Deutschen Krebsgesellschaft zu Vitamin B17 (Amygdalin) https://www.krebsgesellschaft.de/deutsche-krebsgesellschaft/klinische-expertise/wissenschaftliche-stellungnahmen.html. Zugegriffen: 24. Okt. 2018

Jeon T-I, Zhu B, Larson JL, Osborne TF (2008) SREBP-2 regulates gut peptide secretion through intestinal bitter taste receptor signaling in mice. J Clin Invest 118:3693–3700

Kandel C, Schmidt P, Perniss A, Keshavarz M, Scholz P, Osterloh S, Althaus M, Kummer W, Deckmann K (2018) ENaC in cholinergic brusch cells. Front Cell Dev Biol 6. https://doi.org/10.3389/fcell.2018.00089

Kaufmann J (2016) Mythenjagd (1): Bio bedeutet ungespritzt. Salonkolumnisten. http://www.salonkolumnisten.com/mythenjagd-1-bio-bedeutet-ungespitzt/. Zugegriffen: 23. Okt. 2018

Klotter C (2016) Identitätsbildung über Essen. essentials. Springer Fachmedien, Wiesbaden. https://doi.org/10.1007/978-3-658-13309-2_3

Kojima I, Medina J, Nakagawa Y (2017) Role of the glucose-sensing receptor in insulin secretion. Diab Obes Metab 19(Suppl. 1):54–62

Lee RJ, Cohen NA (2014) Bitter and sweet taste receptors in the respiratory epithelium in health and disease. J Mol Med 92:1235–1244

Lewandowski BC, Sukumaran SK, Margolskee RF, Bachmanov AA (2016) Amiloride-insensitive salt taste is mediated by two populations of type III taste cells with distinct transduction mechanisms. J Neurosci 36(6):1942–1953

Lu P, Zhang C-H, Lifshitz LM, ZhuGe R (2017) Extraoral bitter taste receptors in health and disease. J Gen Physiol 149(2):181–197

Martin RA (2010) ReefQuest centre for shark research. http://www.elasmo-research.org/education/white_shark/sensory_bio.htm. Zugegriffen: 24. Okt. 2018

Molteri M (2016) A magical mushroom powder blocks bitterness in food. Wired. https://www.wired.com/2016/08/magical-mushroom-powder-blocks-bitterness-food/. Zugegriffen: 15. Okt. 2018

Neukamm M (2014) Der Evolutionsbeweis in unserem Blut. http://evobioblog.de/der-evolutionsbeweis-unserem-blut/. Zugegriffen: 9. Okt. 2018

Reed DR, Xia MB (2015) Recent advances in fatty acid perception and genetics. Adv Nutr 6:353S–360S

Riedel K, Sombroek D, Fiedler B, Siems K, Krohn M (2017) Human cell-based taste perception – a bittersweet job for industry. Nat Prod Rep 34:484–495

Ritzer J, Miesler T, Meinel L (2017) Bioresponsive Diagnostik – die Zunge als Detektor oraler Entzündungen. Biospektrum 7(17):782–784

Robert Koch-Institut (2017) Magenkrebs (Magenkarzinom). Zentrum für Krebsregisterdaten. https://www.krebsdaten.de/Krebs/DE/Content/Krebsarten/Magenkrebs/magenk rebs_node.html. Zugegriffen: 23. Okt. 2018

Roper SD (2014) TRPs in taste and chemesthesis. Handb Exp Pharmacol 223:827–871

Rozin P, Schiller D (1980) The nature and acquisition of a preference for chili pepper by humans. Motivation and Emotion 4(7):77–101

Schaumburg HH, Byck R, Gerstl R, Mashman JH (1969) Monosodium L-glutamate: its pharmacology and role in the Chinese restaurant syndrome. Science 163:826–828

Schöbel N, Radtke D, Kyereme J, Wollmann N, Cichy A, Obst K, Kallweit K, Kletke O, Minovi A, Dazert S, Wetzel CH, Vogt-Eisele A, Gisselmann G, Ley JP, Bartoshuk LM, Spehr J, Hofmann T, Hatt H (2014) Astringency is a trigeminal sensation that involves the activation of G protein-coupled signaling by phenolic compounds. Chem Senses 39:471–487

Shaik FA, Singha N, Arakawaa M, Duana K, Bhullara RP, Chelikania P (2016) Bitter taste receptors: extraoral roles in pathophysiology. Int J Biochem & Cell Biol 77:197–204

Smallwood K (2016) Do sharks really not Like how humans taste? Today I found out. http://www.todayifoundout.com/index.php/2016/07/sharks-actually-like-taste/. Zugegriffen: 24. Okt. 2018

Starz S (2017) Produkttest: Naturjoghurt im Falstaff-Check. Falstaff 01/2017. https://www.falstaff.at/nd/produkttest-naturjoghurt-im-falstaff-check/. Zugegriffen: 24. Okt. 2018

Steensels S, Depootere I (2018) Chemoreceptors in the gut. Annu Rev Physiol 80:117–141

Ueda Y, Sakaguchi M, Hirayama K, Miyajima R, Kimizuka A (1990) Characteristic flavor constituents in water extract of garlic. Agric Biol Chem 54(1):163–169

Zocchi D, Wennemuth G, Oka Y (2017) The cellular mechanism for water detection in the mammalian taste system. Nature Neurosci 20(7):927–935

Printed in the United States
by Baker & Taylor Publisher Services